W9-DHJ-907

Read All About
Earthly Oddities

WAVES AND TIDES

Patricia Armentrout

The Rourke Press, Inc.
Vero Beach, Florida 32964

© 1996 The Rourke Press, Inc.

All rights reserved. No part of this book may be reproduced or utilized in any form or by any means, electronic or mechanical including photocopying, recording or by any information storage and retrieval system without permission in writing from the publisher.

PHOTO CREDITS
© A/P Wide World Photos: pg. 19; © Dick Dietrich: Cover; © East Coast Studios: pgs. 12, 16; © Warren Faidley/Int'l Stock: pg. 18; © Kim Karpeles: pgs. 13, 21; © NASA: pgs. 4, 15; © James P. Rowan: pg. 10; © Johnny Stockshooter/Int'l Stock: pgs. 7, 9; © Oscar C. Williams: pgs. 6, 22

ACKNOWLEDGMENTS
The author wishes to acknowledge David Armentrout for his contribution in writing this book.

Library of Congress Cataloging-in-Publication Data

Armentrout, Patricia, 1960-
Waves and tides / by Patricia Armentrout.
p. cm. — (Earthly Oddities)
Includes index.
Summary: Discusses the causes of ocean waves and tides and their effects on both people and the Earth.
ISBN 1-57103-157-X
1. Ocean Waves—Juvenile literature. 2. Tides—Juvenile literature. [1. Ocean waves. 2. Tides]
I. Title II. Series: Armentrout, Patricia, 1960- Earthly Oddities.
GC211.6.A76 1996
551.47'02—dc20 96-2882
CIP
AC

Printed in the USA

TABLE OF CONTENTS

WAVES

Over 70 percent of the Earth's surface is covered by water. Most of the water on Earth is in our oceans.

The ocean waters are in constant motion. The sun, wind, and even the moon combine to keep the seas moving. All of this movement is energy. Waves are just one way the ocean transfers or releases its energy.

As wind blows across the ocean, it pushes water in front of it. The energy from the wind is now in the form of waves. Wind-blown waves can travel for hundreds of miles before crashing onto some distant shore.

Over 70 percent of the Earth's surface is covered by water.

WAVES AND PEOPLE

Waves affect people in different ways. Surfers look for the perfect wave to catch a ride on. Swimmers enjoy playing in the ocean surf. Jet skiers ride up and down on the ocean swells.

People all over the world love to play in the ocean surf.

Ships are hard to navigate in ocean waves like these.

Waves may provide a challenge for many types of water sports, but they also affect people in a more serious way. Captains of ships and smaller boats must learn to **navigate** (NAV eh gayt) safely on the ever-changing sea. Homeowners must keep a constant watch on the ocean as big waves can mean big trouble to homes on the water.

Can you think of other ways that waves can affect people?

MONSTER WAVES

The biggest waves are caused by undersea earthquakes and volcanoes. These giant waves are often called tidal waves, although they have nothing to do with tides. The correct name is **tsunami** (tsoo NAH mee).

Tsunamis can travel at speeds of over 400 miles an hour in the open ocean. In deep water, the waves may be only two to four feet high; but as they approach the shallow water near land, they can become monsters.

In 1883 a tsunami struck the Island of Java in the Pacific Ocean. The wave, estimated at over 100 feet high, killed more than thirty-six thousand people.

Giant waves are strong enough to destroy anything in their path.

PREDICTING TSUNAMIS

When an undersea earthquake occurs, it causes vibrations called shock waves. Shock waves push water in front of them creating huge ocean waves, or tsunamis.

By studying undersea earthquakes, scientists can now predict some tsunamis. These scientists are called **seismologists** (syz MAHL eh jists). They use special tools that can detect earthquakes.

The tsunami warning system in Hawaii monitors earthquakes in the Pacific Ocean. When an earthquake is detected, people living in this area are warned of a possible tsunami. People living near shore move to higher ground until the danger passes.

The residents of this coastal home lost part of their house due to high water.

TIDES

Each day the ocean waters move in toward shore, and out away from shore, in a never-ending cycle. This cycle is called the tide. An incoming tide is a flood, or high tide. An outgoing tide is an ebb, or low tide.

Low tide is the best time to find shells on the beach.

High tide can completely cover a beach with waves.

Like the Earth, the sun and moon have gravity. Gravity is the force that keeps you from floating into space. The gravity from the sun and moon pulls the ocean waters toward them, causing the flood and ebb tides.

STORM SURGE

A storm surge is a series of large waves that crash onto coastlines in many parts of the world. The waves are caused by tropical storms called **hurricanes** (HER eh kaynz) and **typhoons** (ty FOONZ).

The storms produce high winds that whip the ocean waters into a froth. As the storm approaches land, it pushes water ahead of it. If a storm comes ashore during a high tide, serious flooding and damage can occur.

Most damage and loss of life from tropical storms comes not from the dangerous winds but from the storm surge.

This picture from space shows the swirling clouds of a typhoon.

THE CHANGING COASTLINE

The next time you go to the ocean, watch as the waves crash onto shore, one right after another. Now imagine this process happening over and over again for thousands of years.

Each time a wave comes ashore it leaves the beach, or coastline, a little different than it was before. This change is **erosion** (eh RO zhun). Over time, erosion can change the look and shape of the coastline.

Sometimes a coastline can change quickly. Powerful storm surges can change the coastline in a single day. Huge walls of water wash away millions of tons of earth as the water drains back into the ocean.

The coastline is shaped in part by waves that constantly crash on shore.

THE AFTERMATH

Great walls of water pound the shore, washing away everything in their path. Homes and businesses are destroyed. Roads and bridges are washed out to sea. A massive clean-up operation is needed. In the United States, government agencies rush to aid people by providing emergency services and supplies.

Hurricanes can destroy homes and businesses.

Supplies continue to flood this center after Hurricane Andrew hit South Florida.

Wildlife is also affected by storms. Most animals seek shelter during a storm and usually recover very quickly. Many plants may be lost but most will grow back over time.

PROTECTION FROM THE STORM

Many cities around the world have been built in low-lying coastal areas. This location puts the people there at risk during tropical storms.

Different methods are used to help protect people from high waters. Some cities build walls of earth along the shore to control the water. The walls are called dikes or floodwalls.

Before new cities are built, studies are done to make sure the area will be safe from dangerous ocean storms. When a storm does occur, advance warning allows people to seek the shelter of higher ground.

Walls are built along the shore to help prevent erosion.

GLOSSARY

erosion (eh RO zhun) — the process of destroying or wearing away

hurricanes (HER eh kaynz) — a rotating wind storm with winds reaching 74 miles an hour or greater

navigate (NAV eh gayt) — to sail or steer a course on a ship or aircraft

seismologists (syz MAHL uh jists) — people who deal with and study earthquakes and vibrations of the Earth

tsunami (tsoo NAH mee) — a wave caused by an earthquake or volcanic eruption

typhoons (ty FOONZ) — rotating wind storms in the China Sea area

Ocean waves are a constant source of natural energy.

INDEX

BOOKS ARE GREAT
[illegible] AT
HAMILTON [illegible] SCHOOL LIBRARY